Pascal Egbenda (Ed.)
Foday Thullah

A Phisico-Chemical Analysis of soil and fruits for Heavey Metals

Pascal Egbenda (Ed.)
Foday Thullah

A Phisico-Chemical Analysis of soil and fruits for Heavey Metals

LAP LAMBERT Academic Publishing

Impressum / Imprint

Bibliografische Information der Deutschen Nationalbibliothek: Die Deutsche Nationalbibliothek verzeichnet diese Publikation in der Deutschen Nationalbibliografie; detaillierte bibliografische Daten sind im Internet über http://dnb.d-nb.de abrufbar.
Alle in diesem Buch genannten Marken und Produktnamen unterliegen warenzeichen-, marken- oder patentrechtlichem Schutz bzw. sind Warenzeichen oder eingetragene Warenzeichen der jeweiligen Inhaber. Die Wiedergabe von Marken, Produktnamen, Gebrauchsnamen, Handelsnamen, Warenbezeichnungen u.s.w. in diesem Werk berechtigt auch ohne besondere Kennzeichnung nicht zu der Annahme, dass solche Namen im Sinne der Warenzeichen- und Markenschutzgesetzgebung als frei zu betrachten wären und daher von jedermann benutzt werden dürften.

Bibliographic information published by the Deutsche Nationalbibliothek: The Deutsche Nationalbibliothek lists this publication in the Deutsche Nationalbibliografie; detailed bibliographic data are available in the Internet at http://dnb.d-nb.de.
Any brand names and product names mentioned in this book are subject to trademark, brand or patent protection and are trademarks or registered trademarks of their respective holders. The use of brand names, product names, common names, trade names, product descriptions etc. even without a particular marking in this work is in no way to be construed to mean that such names may be regarded as unrestricted in respect of trademark and brand protection legislation and could thus be used by anyone.

Coverbild / Cover image: www.ingimage.com

Verlag / Publisher:
LAP LAMBERT Academic Publishing
ist ein Imprint der / is a trademark of
OmniScriptum GmbH & Co. KG
Heinrich-Böcking-Str. 6-8, 66121 Saarbrücken, Deutschland / Germany
Email: info@lap-publishing.com

Herstellung: siehe letzte Seite /
Printed at: see last page
ISBN: 978-3-659-71277-7

TABLE OF CONTENTS

ACKNOWLDGEMENT vi

CHAPTER ONE: INTRODUCTION... 1

    1.1 Aim of Study.. 4

    1.2: Justification of Study.. 4

CHAPTER TWO: LITERATURE REVIEW................................... 5

    2.1: Heavy Metals.. 5

    2.2: Toxicity of heavy metals.. 6

    2.3: Heavy Metals in the Research.. 10

    2.3.1: Lead.. 10

    2.3.2: Chromium..10

    2.3.3: Zinc..12

    2.3.4: Copper.. 13

2.3.5: Arsenic……………………………………………….......... 14

2.4: Heavy Metals accumulation in Fruits…………………….. 16

2.5: Fate of heavy metals in soil and environment…………………. 18

2.6: Translocation Factor (TF)………………………...………... 21

2.7: Effect of pH on metals in soil……………………….…....... 21

CHAPTER THREE: **METHODOLOGY**………………………….. 25

3.1: Description of sampling site……………………………….. 25

3.2: Sample collection…………………………………….…… 26

3.3: Sample Preparation…………………………………….…... 26

3.3.1: Preparation of Soil samples………………………………..26

3.3.2: Preparation of Fruit samples…………………………….... 27

3.3.3: Digestion of soil samples……………………………........ 27

3.3.4: Digestion of fruit samples……………………………….. 28

3.4: Principles of instrumentation……………………………… 28

3.4.1: Measurement of pH ……………………………………… 28

3.4.2: Conductivity Measurement……………………………... 29

3.4.3: Atomic Absorption Spectrophotometers............................30

3.4.4: Principles of spectrophotometer................................31

3.5: Determination of concentration of metals in soil and fruit samples...34

3.6: Method Detection Limit (MDL)............................. 34

3.7: Preparation of calibration curve using standard compounds......... 34

CHAPTER FOUR: RESULTS............................ 35

4.1. Physicochemical Analysis of the Soil Samples........................ 35

4.1.2: Heavy Metal Concentration in soil and Fruits....................... 36

4.1.3: Calculations.. 39

4.1.4: Translocation Factor (TF)..............................39

CHAPTER FIVE: DISCUSSION OF RESULTS.......................41

5.1: pH and Conductivity analysis in the Soil samples........................41

5.2: Determination of Heavy Metals Concentration in the Soil and in the Fruits....41

5.3: Translocation Factor (TF) Between Soil and Fruits..........................42

5.4: Correlation Coefficient (r^2) Between Soil and Fruits........................43

CHAPTER SIX:

CONCLUSION... 47

REFRENCES.. 48

LIST OF TABLES

Table 1. Fruit samples analyzed.. 3

Table 2. Typical total concentration of Fe, Cu, Zn and Mn in soil.................. 23

Table 3. pH and conductivity results.. 35

Table 4. Recommended values for pH and Conductivity by WHO and EU..........36

Table 5 Concentration of dissolved metals in soil and accumulation in fruits at

Mokaba rehabilitated mined site (in mg/L) dry weight............................... 37

Table 6. Recommended limits of investigated metals (Pb, Zn, Cr, Cu and As by

World Health Organization (WHO) in mg/L... 38

Table 7. Comparison of heavy metals concentration in soil and accumulation in the

fruits of Mokaba rehabilitated mined site (in mg/L).................................... 39

Table 8: Heavy Metals (HMs) Concentration in Soil (CS), Accumulation in Mango

Fruit (AMF), Accumulation in Cashew Fruit (ACF) and their Translocation Factors

(TF) in mango fruit and cashew fruit respectively...................................40

Table 9. Estimation of correlation coefficient between soil and mango fruit.........44

Table 10. Estimation of correlation coefficient between soil and cashew fruit.......45

LIST OF FIGURES

Figure 1. A Sketched Map of the Sample Collection Site..............................26

Figure 2. Atomic Absorption Spectrometer Block Diagram............................33

Figure 3. Translocation Factor of Heavy Metals from Soil to Mango Fruit...........40

Figure 4 Translocation Factor of Heavy Metals from Soil to the Cashew Fruit...... 40

ACKNOWLEDGEMENT

We thank God for giving us the inspiration to carry out this work.

We also thank our colleagues in the Department of Chemistry, Fourah Bay College University of Sierra Leone for various pieces of advice we received from them during the experimental stage as well as during preparation of the manuscript.

The contribution to this work from Mr. Ibrahim Kamara, our former final year Honours student, who carried out the research under our supervision, cannot be forgotten.

We also wish to commend our families for their understanding and cooperation during our venture.

Thanks also to the University of Sierra Leone for creating the conducive atmosphere under which we worked.

CHAPTER ONE

INTRODUCTION

Heavy metals are significant environmental pollutants, and their toxicity is a problem of increasing significance for ecological, evolutionary, nutritional and environmental reasons. The term "heavy metals" refers to any metallic element that has a relatively high density and is toxic or poisonous even at low concentration. "Heavy metals" is a general collective term, which applies to the group of metals and metalloids with density greater than 4 g/cm^3 or 5 times or more than water. However, the chemical properties of the heavy metals are the most influencing factors compared to their density (Hawkes, 1997).

Environment is defined as the totality of circumstances surrounding an organism or group of organisms such as, the combination of external physical conditions that affect and influence the growth, development and survival of organisms (Farlex Incorporated, 2005). The environment is considered in food, and the less tangible, the communities we live in. A pollutant is any substance in the environment, which causes objectionable desire to the environment or is a substance or energy introduced into the environment that has undesired effects, or adversely affects the usefulness of a resource (Wikipedia, the free encyclopedia, 2013).

The effect of heavy metal contamination of fruit and vegetables cannot be underestimated as these foodstuffs are important components of human diet. Fruit and vegetables are rich sources of vitamins, minerals and fibers and also have beneficial

1

anti-oxidative effects. However, the intake of fruits and vegetables contaminated with heavy metals may pose a risk to human health. Hence, the analysis of foods to ascertain whether or not they carry heavy metal contaminants which is one of the most important aspects of food quality assurance. Heavy metals, in general, are not biodegradable, have long biological half-lives, and have the potential for accumulation in different body organs, leading to unwanted side effects. Plants take up heavy metals by absorbing them from airborne deposits on the parts of the plants exposed to the air or from contaminated soils through root systems. Also, the heavy metal contamination of fruit and vegetables may occur through irrigation with contaminated water (M. Sal Jasir et al., 2005).

Generally heavy metals originate from two primary sources: natural inputs (e.g. parent material weathering) and anthropogenic inputs (metalliferous industries and mining, vehicle exhaust, agronomic practices, etc). In recent decades, the natural input of heavy metals due to pedogenic processes has been greatly exceeded by human input, and the human sources have also become more complex (Facchielli et al., 2001).

Excessive accumulation of heavy metals in agricultural soils results in soil contamination and has consequences for food quality and safety. Food is the major intake source of toxic metals by human beings. Among food system, vegetables and fruits are the most exposed food to environmental pollution due to aerial burden. They take up heavy metals and accumulate them in their edible and non-edible parts at quantities high enough to cause clinical problems to both animals and human beings. Excessive content of metals beyond Maximum Permissible Level (MPL) leads to a number of nervous, cardiovascular, renal, neurological impairment as well as bone diseases and several other health disorders (WHO, 1992). Vegetables and fruits are an essential part of diet and are taken both cooked and raw forms by human. They act as buffering agents for acid generation during digestion (Maleki and Zarasvand 2008) and some metals present in fruits and vegetables are even important biochemically and psychologically from health point of view. Metals like cobalt (Co), chromium (Cr), copper (Cu), iron (Fe), manganese (Mn), molybdenum (Mo), selenium (Se) and zinc (Zn) help in regulating human metabolism. Manganese is an essential element which acts as an activator and constituent of many enzymes present in human (Sresty and Rao 1999). But some elements like Pb,Cd and As are very toxic for human. Other elements such as Cr, Co, and Ni although essential for human but at concentrations higher than those recommended may cause metabolic disorders.

Table 1 Fruit samples analyzed

Common name	Designation	Scientific name	Edible part
Cashew	Cas	Anacardium occidentale L	Fruit
Mango	Man	Mangifera indica L	Fruit

1.1: Aim of study

In this research the aim is to use spectrophotometric methods such as Flame Atomic Absorption Spectrophotometry (FAAS) to determine heavy metal concentration in soil and in the selected fruit samples such as cashew nut (anacardium occidentale) and mango fruit (Mangifera indica L) collected from rehabilitated mined out areas in Mokaba, in the Sierra Rutile environs.

1.2: Justification of study

Most local communities in Sierra Leone largely consume mango fruit (Mangifera indica L) and to some extent cashew fruit (Anacardium Occidentale) as food supplement. The mining activities carried out by the Sierra Rutile mining company may have exposed some of the heavy metals onto the soil surface. It is therefore highly likely that any crops planted on the rehabilitated lands will accumulate some of the metals over time. This study could be used to ascertain soil pollution and as well as provide guidance for pollution assessment and control in the rehabilitated lands in the Sierra Rutile environs.

CHAPTER TWO

LITERATURE REVIEW

2.1: Heavy Metals

Heavy metals are elements having atomic weight between 63.545 and 200.5 g and a specific gravity greater than four (Bala M, Shehu R.A, Lawal, 2008). The elements play essential roles in biological processes, but at higher concentrations they may be toxic to the biota and they disturb the biochemical processes and cause hazards. These elements include metals (Cd, Cr, Co, Pb, Cu, Zn, Pd, Ni, and Ag) and metalloids (Se, As, Sb). Most of the trace elements are transition metals with variable oxidation states and coordination numbers.

These metals form complexes with organics in the environment thereby increasing their mobility in the biota and manifest toxic effect (Ekpete OA et al., 2010).

Heavy metals have been reported to have positive and negative roles in human life. Some like Cd, Pb and Hg are major contaminants of food supply and may be considered the most important problem to our environment while others like Fe, Zn and Cu are essential for biochemical reactions in the body (Sobukola OP et al., 2010).

Intracellular free magnesium is involved in the energy reactions of phosphorylation and is necessary for the activation of hundreds of enzymatic reactions concerning adenosine-5'-triphosphate (ATP). (Dube L, Granry JC, 2003). Ca is an essential nutrient required for critical biological functions such as nerve conduction, muscle contraction, cell adhesiveness, mitosis, blood coagulation and structural support of the skeleton (Miller GD, Jarvis JK, McBean LD., 2001).

Generally, most heavy metals are not biodegradable and have long biological half-lives and they have the potential for accumulation in different body organs leading to unwanted side effects. The content of essential elements in plants is conditional, being affected by the characteristics of the soil and the ability of plants to selectively accumulate some metals. Several plants are widely used widespread for their many therapeutic and pharmaceutical virtues, especially anti-oxidant and anti-infectious activities. A big part of the world's population still relies on the benefits of food for the treatment of common illnesses. Food chain contamination by heavy metals has become a burning issue in recent years because of their potential accumulation in bio systems through contaminated water, soil and air (Mahdavian SE, Somashekar RK., 2008).

Due to their persistent nature and cumulative behavior as well as potential toxicity effects, heavy metals absorbed in fruits and vegetable must be investigated to ensure that their levels meet the agreed international requirements.

2.2: Toxicity of heavy metals

Heavy metal toxicity is a major problem of sour environment and they are also one of the major contaminants of our food supply. This problem is receiving more and more attention all over the world, in general and in developing countries in particular. The biological half-lives of these heavy metals are long and have potential to accumulate in different body organs and thus produce unwanted side effects (Jarup amd Ata et al., 2003 and 2009 respectively).

Lead and Cadmium are the most toxic and the most abundant metals in food. Excessive accumulation of these heavy metals in humanan bodies creates problems like cardiovascular, kidney, nervous and bone diseases (Jarup and Ata et al., 2003, WHO., 1992.)

Recently pollution of the general environment has generated an increase global interest. In this respect, contamination of agricultural soils with heavy metals has always been considered a critical challenge in the scientific community (Faruk et al., 2006). Heavy metals are generally present in agricultural soils at low levels. Due to their cumulative behaviour and toxicity, however, they have a potentially hazardous effect not only on crops but also on human health. Even metals essential to plant growth, like copper (Cu), manganese (Mn), molybdenum (Mo), and zinc (Zn) can be toxic at high concentrations in the soil. Some elements not known to be essential to plant growth, such as arsenic (As), barium (Ba), cadmium (Cd), chromium (Cr), lead (Pb), nickel (Ni), and selenium (Se), also are toxic at high concentrations or under certain environmental conditions in the soil (Slagle et al., 2004). Both pH and redox potential can affect the toxicity of heavy metals by limiting their availability. At low pH, metals generally exist as free cations; at alkaline pH, however, they tend to precipitate as insoluble hydroxides, oxides, carbonates, or phosphates (Mamboya, F.A., 2007). Chemical hazards include chemical agents such as heavy metals, nutrients such as nitrogenous compounds, phosphorus compounds, minerals, insecticides, pesticides, fertilizers, fungicides, herbicides and organic hazards (Nabulo et al., 2008).

Metals, unlike the hazardous organics cannot be degraded. Some metals such as Cr, As, Se, and Hg can be transformed to other oxidation states in soil, thus influencing their mobility and toxicity (McLean et al., 1992). Many of them (Hg, Cd, Ni, Pb, Cu, Zn, Cr, Co) are highly toxic both in elemental and soluble salt forms. High concentration of heavy metals in soils is toxic for soil organisms: bacteria, fungi and higher organisms (Woolhouse et al., 1993). Short-term and long-term effects of pollution differ depending on metal and soil characters.

In the after-effect of heavy metal pollutions, the role of pollutant bounding or leaching increases which determines their bioavailability and toxicity. When soil is acidified it increases the concentration of free aluminium ions in the water that is in the soil, and these are potentially toxic to the root systems of plants.

The mobility of many heavy metals also increases when soil becomes more acidic. Perhaps the most serious consequence of the higher metal concentrations is their negative effect on many of the decomposers that live in the soil (Elvingson et al., 2004). The U.S. Environmental Protection Agency (U.S. EPA., 1993) regulates nine trace elements for land-applied sewage sludge: As, Cd, Cu, Pb, Hg, Mo, Ni, Se and Zn. Only six of these elements (Cu, Ni, Zn, Cd, Pb, Se) are considered to be phytotoxic.

Other than the natural source of environmental lead, zinc, cadmium, nickel and copper pollution, the mining industries of these elements should be a potential source of releasing these metals into the atmosphere. Pb in the metallic state and compounds has been used in many ways including the following; as a pigment and drier for some paints, in batteries, in water pipes, as putty with linseed oil for joints in pipes and plates, and in the petro-industry (Brown. G.I., 1983).

Tetraethyl lead used to be added to petrol to improve the octane rating of fuel by preventing premature explosions. Ethylene dibromide is added to leaded gasoline to convert the lead oxide formed in combustion to volatile lead bromide. (Roberts J.D., 1964 and Caserio M.C., 1974). Combustion of leaded petrol is the greatest single contribution to airborne lead. (The Royal Society of Chemistry., June, 1982). In areas where the traffic is high, the amount of lead in the atmosphere is anticipated to form an appreciable percentage.

Like any other elements, Ni, Pb, Cd, Cu and Zn may enter the body via three major routes; inhalation from the atmosphere, absorption from the gastro-intestinal tract and absorption through the skin. Inhalation from the atmosphere can occur from combustion of leaded petrol, burning off of lead paint and other major sources of environmental pollution. Absorption from the gastro-intestinal tract of a significant amount of the body burden of these elements is from food and water through the intestine. Skin absorption has been little studied and probably is of little importance.

2.3: Heavy Metals in the Research

2.3.1: Lead

Lead has been shown to have toxic impact on a variety of metabolic processes essential to plant growth and development, including photosynthesis, transpiration, DNA synthesis, and mitotic activity (Wierzbicka, M., 1999).

Sources of lead include metal smelting, pigments, lead battery manufacturing and leaded petrol. In soil Lead tightly binds itself to organic soil particles which may decrease the mobility of lead in most soils and may reduce uptake by plants (Cooper et al., 1999). It has been suggested that the mobility of lead and copper is greater in sandy soils, which tend to lack organic matter, than in organic soils. Lead has two quite distinct toxic effects on human beings, physiological and neurological. "The relatively immediate effects of acute lead poisoning are ill defined symptoms, which include nausea, vomiting, abdominal pains, anorexia, constipation, insomnia, anemia, irritability, mood disturbances and coordination loss. In more severe situations neurological effects such as restlessness, hyperactivity, confusion and impairment of memory can result as well as coma and death" (Ansari, et al. 2004).

2.3.2: Chromium

Chromium exists in two possible oxidation states in soils: the trivalent chromium, Cr (III) and the hexavalent chromium, Cr(VI). Forms of Cr (VI) in soils are as chromate ion, $HCrO_4^-$ predominant at pH<6.5, or Cr(VI), predominant at pH 6.5, and as dichromate, $Cr_2O_7^{2-}$ predominant at higher concentrations (>10mM) and at pH 2-6. The dichromate ions pose a greater health hazard than chromate ions. Cr (VI) ions are more toxic than Cr (III) ions. Reviews of the processes that control the fate of chromium in soil and the effect these processes have on remediation are given in. Because of the anionic nature of Cr (VI), its association with soil surfaces is limited to positively charged exchange sites, the number of which decreases with increasing soil pH. Iron and aluminium oxide surfaces will adsorb Cr(VI) at acidic and neutral pH. Adsorption of Cr (VI) by ground-water alluvium was due to the iron oxides and hydroxides coating the alluvial particles. The adsorbed Cr (VI) was, however, easily desorbed with the input of uncontaminated ground water, indicating nonspecific adsorption of Cr (VI).

The presence of chloride and nitrate had little effect on Cr (VI) adsorption, whereas sulfate and phosphate inhibited adsorption. SO_4^{2-} and dissolved inorganic carbon inhibited Cr (VI) adsorption by amorphous iron oxyhydroxide and subsurface soils. The presence of sulfate, however, enhanced Cr (VI) adsorption to kaolinite (Zachara et al., 1988). BaCrO4 may form in soils at chromium contaminated waste sites.

No other precipitates of hexavalent compounds of chromium have been observed in a pH range of 1.0 to 9.0 (Hexavalent chromium is highly mobile in soils. In a study of the relative mobilities of 11 different trace metals in a wide range of soils, clay soil, containing free iron and manganese oxides, significantly retarded Cr (VI) migration. Hexavalent chromium was found to be the only metal studied that was highly mobile in alkaline soils (Griffin et al., 1978). The parameters that correlated with Cr (VI) immobilization in the soils were free iron oxides, total manganese, and soil pH. On the other hand, soil properties, cation exchange capacity, surface area, and percent clay had no significant influence on Cr (VI) mobility. Cr(III) forms hydroxy complexes in natural water, including $Cr(OH)^{2+}$, $Cr(OH)^{2+}$, $Cr(OH)_3 0$, and $Cr(OH)^{4-}$ (Rai et al., 1987).

Trivalent chromium is readily adsorbed by soils. In a study of the relative mobility of metals in soils at pH 5, Cr (III) was found to be the least mobile. Hydroxy species of Cr(III) precipitate at pH 4.5 and complete precipitation of the hydroxy species occurs at pH 5.5. Hexavalent chromium can be reduced to Cr (III) under normal soil pH and redox conditions. Soil organic matter has been identified as the electron donor in this reaction (Bartlett et al., 1976).

Industrial use of chromium also includes organic complexes of Cr (III). Chromium (III) complexed with soluble organic ligands will remain in the soil solution (James and Bartlett, 1983a). In addition to decreased Cr(III) adsorption, added organic matter also may facilitate oxidation of Cr(III) to Cr(VI).

2.2.3: Zinc

Zinc is readily absorbed by clay minerals, carbonates, or hydrous oxides. The greatest percent of the total Zn in polluted soils and sediments is associated with Fe and Mn oxides. Precipitation is not a major mechanism of retention of Zn in soils because of

12

the relatively high solubility of Zn compounds. Precipitation may become a more important mechanism of Zn retention in soil-waste systems. As with all cationic metals, Zn adsorption increases with pH. Zinc hydrolysizes at pH>7.7 and these hydrolyzed species are strongly adsorbed to soil surfaces. Zinc forms complexes with inorganic and organic ligands that will affect its adsorption reactions with the soil surface (Hickey et al., 1984). Zinc has been well known to be an important trace element as a cofactor for insulin. Zinc is a very common environmental contaminant and usually outranks all other metals and it is commonly found in association with lead and cadmium (Finkelman, 2005).

Acidic soils and sandy soils with a low organic content have a reduced capacity for zinc absorption. In neutral to alkaline soils, $Zn (OH)^+$ is a dominant solution species of Zn that may adsorb to soil and replace one H^+. The pH value at which the change in the mechanism that controls solubility occurs will depend on the soil properties. At low pH the total soluble Zn and free Zn activity are nearly the same, but at high pH they are not the same because of hydrolysis species, organic Zn species, etc. (Catlett et al., 2002).

2.3.4 : Copper

Copper is retained in soils through exchange and specific adsorption mechanisms. At concentrations typically found in native soils, Cu precipitates are unstable. This may not be the case in waste-soil systems and precipitation may be an important mechanism of retention. Clay mineral exchange phase may serve as a sink for Cu in noncalcareous

13

soils (Cavallaro and McBride., 1978). In calcareous soils, specific adsorption of Cu onto CaCO3 surfaces may control Cu concentration in solution (Cavallaro et al., 1978).

Cu may exist in soils in the following forms: (i) water soluble, (ii) exchangeable, (iii) organically bound, (iv) associated with carbonates and hydrous oxides of Fe, Mn, and Al, and (v) residual. Copper is adsorbed into the soil, forming an association with organic matter, Fe and Mn oxides, soil minerals, etc., thus making it one of the least mobile of the trace metals (Ioannou et al., 2003).

It is an essential enzymatic element. Certain concentrations of copper are necessary for normal biological activities of amino oxides and tyrosinase enzyme. Tyrosinase is the enzyme that is required for catalytic conversion of tyrosine to melanin, which is a vital pigment located beneath the skin, and thus protects the skin from dangerous radiations (Hashmi, D. R. et al., 2007).

2.3.5: Arsenic

In the soil environment arsenic exists as either arsenate, ($AsO4^{3-}$), or as arsenite, (AsO^2). Arsenite is the more toxic form of arsenic.

The behavior of arsenate in soil is analogous to that of phosphate, because of their chemical similarity. Like phosphate, arsenate forms insoluble precipitates with iron, aluminum, and calcium. Iron in soils is most effective in controlling arsenate's mobility. Arsenite compounds are reported to be 4-10 times more soluble than arsenate compounds. In the adsorption by kaolinite and montmorillonite, maximum adsorption of As(V) occur at pH 5. Adsorption of arsenate by aluminum and iron oxides has shown an adsorption maximum at pH 3-4 followed by a gradual decrease in adsorption with increasing pH (Anderson et al., 1976). The mechanism of adsorption has been ascribed to inner sphere complexation (specific adsorption), which is the same mechanism controlling the adsorption of phosphate by oxide surfaces. The adsorption of arsenite, As(III), is also strongly pH- dependent. It was observed that an increase in sorption of As (III) by kaolinite and montmorillonite occur over a pH range of 3-9 (Griffin, Shimp, 1978). The maximum adsorption of As (III) by iron oxide occurred at pH 7. Adsorption of As (III) is rapid and irreversible on some soils.

Both pH and redox are important in assessing the fate of arsenic in soil. At high redox levels, As(V) predominates and arsenic mobility is low. As the pH increases or the redox decreases As (III) predominates. The reduced form of arsenic is more subject to leaching because of its high solubility. The reduction kinetics are, however, slow. Formation of As (III) also may lead to the volatilization of arsine (AsH_3) and methyl-arsines from soils (Woolson et al., 1977).

Under soil conditions of high organic matter, warm temperatures, adequate moisture, and other conditions conducive to microbial activity, the reaction sequence is driven towards methylation and volatilization. Only 1 to 2 percent of sodium arsenate applied at a rate of 10 ppm is found to be volatilized in 160 days (Woolson et al., 1977). The loss of organic arsenic compounds from the soil was far greater than for the inorganic source of arsenic. Arsenite, As(III), can be oxidized to As(V). Manganese oxides are the primary electron acceptor in this oxidation (Oscarson et al., 1983).

2.4: Heavy Metals accumulation in Fruits

Heavy metals are of interest due to their abundance in the environment, which has increased considerably as a result of human activities. Their fate in polluted soils is a subject of study because of the direct potential toxicity to biota and the indirect threat to human health via the contamination of groundwater and accumulation in food crops (Martinez and Motto, 2000).

Heavy metals are dangerous because they tend to bioaccumulate. This means that the concentration of a chemical in a biological organism becomes higher relative to the environmental concentration. Heavy metal pollution of soil enhances plant uptake causing accumulation in plant tissues and eventual phytotoxicity and change of plant community (GimmlerI et al., 2002).

In environments with high nutrient levels, metal uptake can be inhibited because of complex formation between nutrient and metal ions. Therefore, a better understanding of heavy metal sources, their accumulation in the soil and the effect of their presence in water and soil on plant systems seems to be a particularly important issue.

The soil to plant transfer factor is one of the important parameters used to estimate the possible accumulation of toxic elements, especially radionuclides through food ingestion (El-Ghawi et al., 2005). Several studies have indicated that crops grown on soils contaminated with heavy metals have higher concentrations of heavy metals than those grown on uncontaminated soil (Nabulo, 2006). Heavy metals accumulating in soil directly (or through plants indirectly) enter food chains, thus endangering herbivores, indirectly carnivores and not least the top consumer humans. Compounds accumulate in living organisms any time they are taken up faster than they are broken down (metabolized) or excreted (O'Brien, 2008).

Most of the soils in the mined out areas of Sierra Rutile (Mokaba) are sandy, and due to their sandy nature, it is believed that, the company, during its rehabilitation process of the land for planting of food crops and economic trees, they may normally placed in abundant organic fertilizers for healthy growth of the plants. However, heavy metals are believed to be always associated with organic fertilizers, in that regard, this may impose an addition of the heavy metals to the already exposed ones as a result of the mining activities being carried out.

Metals such as lead, arsenic, cadmium, copper, zinc, nickel, and mercury are continuously being exposed to our soils surface through various human activities such as vehicle exhausts and mining, together with anthropogenic sources. All these sources cause accumulation of metals and metalloids in our agricultural soils and pose threat to food safety issues and potential health risks due to soil to plant transfer of metals (El-Ghawi et al., 2005).

2.5: Fate of heavy metals in soil and environment

In soil, metals are found in one or more of several "pools" of the soil:

1) dissolved in the soil solution;

2) occupying exchange sites on inorganic soil constituents;

3) specifically adsorbed on inorganic soil constituents;

4) associated with insoluble soil organic matter;

5) precipitated as pure or mixed solids;

6) present in the structure of secondary minerals; and/or

7) present in the structure of primary minerals.

Metals in the soil solution are subject to mass transfer out of the system by leaching to ground water, plant uptake, or volatilization, a potentially important mechanism for Hg, Se, and As. At the same time metals participate in chemical reactions with the soil solid phase. The concentration of metals in the soil solution, at any given time, is governed by a number of interrelated processes, including inorganic and organic complexation, oxidation-reduction reactions, precipitation/dissolution reactions, and adsorption/desorption reactions.

The ability to predict the concentration of a given metal in the soil solution depends on the accuracy with which the multiphase equilibria can be determined or calculated (Shuman, 1991). The incidence of heavy metal contamination from both natural and anthropogenic sources has increased concern about possible health effects. Natural and anthropogenic sources of soil contamination are widespread and variable (Tahir et al., 2007). Heavy metal contamination of soil results from anthropogenic processes such as mining, smelting procedures and agriculture as-well as natural activities. Heavy metals which are commonly found in high concentrations in landfill leachate are also a potential source of pollution for groundwater (Aziz et al., 2004). Large areas of agricultural land are contaminated by heavy metals that mainly originate from former or current mining activities, industrial emissions or the application of sewage sludge.

Metals exist in one of four forms in the soil: mineral, organic, sorbed (bound to soil), or dissolved. Sorbed metals represent the third largest pool, and are generally very tightly bound to soil surfaces. Although mineral, organic, and sorbed metals are not immediately absorbed by plants, they can slowly release metals into solution.

The inability to determine metal species in soils hampers efforts to understand the mobility, bioavailability and fate of contaminant metals in environmental systems together with the assessment of the health risks posed by them, and the development of methods to remediate metal contaminated sites. However, in some natural soils developed from metal-rich parent materials, as-well as in contaminated soils, up to 30 to 60% of heavy metals can occur in easily unstable forms.

Heavy metals naturally occur in the environment, but may also be introduced as a result of land use activities.

Natural and anthropogenically introduced concentrations of metals in near-surface soil can vary significantly due to different physical and chemical processes operating within soils across geographic regions. Migration of metals in the soil is influenced by physical and chemical characteristics of each specific metal and by several environmental factors.

The most significant environmental factors appear to be (i) soil type, (ii) total organic content, (iii) redox potential, and (iv) pH (Murray et al., 1999). Although heavy metals are generally considered to be relatively immobile in most soils, their mobility in certain contaminated soils may exceed ordinary rates and pose a significant threat to water quality. Organic manure, municipal waste, and some fungicides often contain fairly high concentrations of heavy metals. Soils receiving repeated applications of organic manures, fungicides, and pesticides have exhibited high concentrations of extractable heavy metals and increased concentrations of heavy metals in runoff (Moore et al., 1998). Previous studies indicate that metal constituents of surface soil directly influence the movement of metals, especially in sandy soils (Moore et al., 1998).

2.6: Translocation Factor (TF)

This is the transfer capability of heavy metals from soil to the edible part of fruits.

TF of heavy metals depends upon bioavailability of metals, which in turn depends upon its concentration in the soil, their chemical forms, difference in uptake capability and growth rate of different plant species (FAO/WHO, 2011). Higher values of TF suggest poor retention of metals in soil and/or more translocation in plants (indicates the stronger accumulation of the respective metal by that fruit). The higher uptake of heavy metals in fruits may be due to higher transpiration rate to maintain the growth and moisture content of these plants (Gildon et al., 1981). Also a related study reported highest translocation factor for heavy metals through leafy vegetables. The TF does not present the risk associated with the metal in any form. The degree of toxicity of heavy metal to human beings depends upon their daily intake (N. Sridhara Chary et al., 2008).

2.7: Effect of pH on metals in soil

The link between soil pH and heavy metal threshold values reflects the complex interaction between heavy metals and the various soil properties. pH is a measure of the hydrogen ion concentration acidity or alkalinity of the soil. Measured on a logarithmic scale, a soil at pH 4 is 10 times more acidic than a soil at pH 5 and 100 times more acidic than a soil at pH 6.

Alkalinity is usually an inherent characteristic of soils, although irrigation can increase the alkalinity of saline soils. Soils made alkaline by calcium carbonate alone rarely have pH values above 8.5 and are termed 'calcareous'. Under normal conditions the most desirable pH range for mineral soil is 6.0 to 7.0 and 5.0 to 5.5 for organic soil.

The buffer pH is a value used for determining the amount of lime to apply on acidic soils with a pH less than 6.6. Colloid and metal mobility, was enhanced by decreases in solution pH and colloid size, and increases in organic matter, which resulted in higher elution of sorbed and soluble metal loads through metal–organic complex formation (Karathansis et al., 2005).

Soil weathering often involves soil acidification, and most chemical immobilization reactions are pH dependent. Several studies have demonstrated that pH is the most influential factor controlling sorption–desorption of heavy metals in soils.

The total metal content and soil pH in surface soils are the dominant factors influencing metals content in plants. Measurement of the total concentration of metals in soils is useful for determining the vertical and horizontal extent of contamination and for measuring any net change (leaching to ground water, surface runoff, erosion) in soil metal concentration over time. However, the methods do not give an indication to the chemical form of the metal in the soil system (Mclean and Bledsoe, 1992). Total concentrations of metals in soils are generally a poor indicator of metal toxicity because metals exist in different solution and solid-phase forms that can vary greatly in terms of their bioavailability. Minerals, metals or metalloids, toxic or essential are present in soils in various forms with varying bioavailability, toxicity and mobility.

Most notably, the available Fe concentration represents less than 0.1% of the average total Fe concentration in soils (Jones and Jacobsen, 2003).

Table 2 Typical total concentration of Fe, Cu, Zn and Mn in soil (Lindsay, 1979)

Metal	Average total concentration (mg/kg)
Iron	38,000
Copper	30
Zinc	50
Manganese	600

For all elements, the amounts extracted had a high dependence on pH, and the water solubility of all elements increased sharply with decreasing pH. The pH values at which a sharp increase in element concentration occurred were 4 to 5 for Cu and Zn. At pH 5.0 to 8.0, the solubility of heavy metals, including Cd, Co, Cr, Pb, Zn, and Ni, were generally low, and the released percentages of Cd, Co, Cr, Cu, Ni, Pb, and Zn at pH 5 to 8 were 3.2, 1.4, 0.3, 2.9, 2.8, 1.4, and 2.5%, respectively. Dissolution of Fe and Mn compounds, which plays a role in contaminant mobility in the soil, increases at low p^H. Due to increased metal complexation at high pH, total extractable fractions of metals generally increased (Zhang et al. (2004). Soil reduction mainly depends on four factors: (1) saturation with water and depletion of O2, (2) presence of microorganisms, (3) food for microorganisms, and (4) suitable temperatures. Due to their high degree of reactivity, Mn oxides in soil systems may exert a greater influence on trace metal chemistry (Negra et al., 2005a).

CHAPTER THREE

METHODOLOGY

3.1: Description of sampling site

The Mokaba rehabilitated land is one of the mined out sites that had been rehabilitated several years before the war. It has economic trees such as mango trees, cashew nut trees among others. Fruits from these plants are harvested and sold to communities in the Sierra Rutile environs. The Mokaba site is about half a mile from Mokaba town, a fairly large settlement in the Sierra Rutile environs in the Imperei Chiefdom, Bonthe District, and Southern Sierra Leone.

Mokaba Township has an estimated terrain elevation above sea level of 37 meters, latitude 7° 40′ 3′and longitude 12° 16′ 32.99′ (Google. 2013).

This land was quit productive and healthy. Now the nature of the soil has changed to sandy because of the mining explorations carried out by the Sierra Rutile mining Company.

However, in its attempt to restore these mined out areas, the company, in 2013 rehabilitated 292 hectares of land for useful purposes for affected communities as a way to ignite economic activities especially with the planting of varieties of economic trees and restocking of fish in the lakes for local consumption and marketing in other mined out lands like Gbojeima.

3.2: Sample collection

Soil samples in the site were collected at different points and at variable depths

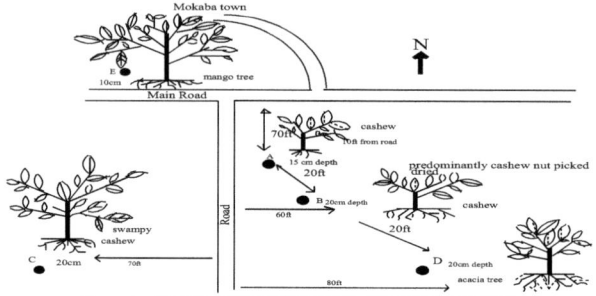

Figure 1. SKETCHED MAP OF THE SAMPLES COLLECTION SITES

ranging from 0—20cm. Fresh ripe fruits of cashew nuts (anacardium occidentale) and mango fruits (Mangifera indica L) were collected from Mokaba rehabilitated lands (see fig.1 above) in the Sierra Rutile Mining environs in April, 2014.

3.3: Sample Preparation

3.3.1: Preparation of Soil sample

The soil samples after air drying at room temperature were sieved with nylon mesh (2 mm). After sieving, the soil sample was ground in an agate and pestle and passed through a 60 micron sieve and then mixed thoroughly to give a representative sample. The soil pH and conductivity were measured in suspension of 1:2 soil to water ratio using calibrated pH and conductivity meters.

After determination of pH and conductivity, the soil samples were then analyzed for heavy metals. These samples were first digested in aqua regia (HCl: HNO_3= 1:3).

After filtration, the solutions were diluted with distilled water and then kept for further analysis.

3.3.2: Preparation of Fruit samples

The cashew fruits (anacardium occidentale) and mango fruits (Mangifera indica L) were washed thoroughly with distilled water. The samples were peeled and sliced using a plastic knife and then dried in oven at 60^0 C. The dried samples were then ground into fine powder and stored in clean stoppered rubber bottles.

3.3.3: Digestion of soil samples

5g of soil samples was accurately weighed and placed in a 250ml beaker. 50ml distilled water was added followed by 50ml aqua regia (45ml HNO_3 and 15ml HCl i.e. 3:1). The solution was mixed and three boiling chips added to prevent vigorous boiling and then heated on a hot plate at 100^0 C for one hour. The mixture was heated for a further 15 minutes at 125^0 C to concentrate it to 5ml volume. The 5ml volume concentrate was removed from the hot plate and allowed to cool. After cooling, a 1ml 30% H_2O_2 was accurately added and then heated for a further 10 minutes. This resulting solution is again cooled. 3ml 30% H_2O_2 was added to the cooled solution and then heated for 10 minutes. 50ml distilled water was added followed by 25ml HCl, mixed and heated to boiling. The resulting hot solution was cooled and then filtered using a Whiteman no.42 filter paper in a volumetric flask and then diluted to 250ml with distilled water.

3.3.4: Digestion of fruit samples

2g of the powdered sample of cashew fruit (anacarduim occidentale) and mango fruit (Mangifera indica L) were placed in separate 100ml beakers. To each beaker 10ml HNO_3 was added and the mixture heated on hot plate at 40^0C for 15 minutes.

The digest was cooled and a further 5ml concentrated HNO_3 added.

The resulting solution was heated for another 30 minutes at 40^0C to concentrate to 5ml volume without boiling. The solutions were cooled. 5ml H_2O was added to each followed by 3ml 30% H_2O_2. The beakers were covered and the solutions gently heated to start the peroxide reaction. The solutions were removed from the hot plate when vigorous effervescence occurs. 1ml 30% H_2O_2 was added to the solutions followed by gentle heating until effervescence subsides. 5ml concentrated HCl was added followed by 10ml distilled water. The resulting solutions were heated for additional 15 minutes without boiling. Finally the solutions were cooled, filtered using a Whiteman no. 42 ash less filter paper and then diluted to 60ml with distilled water.

3.4: Principles of instrumentation

3.4.1: Measurement of pH

The pH meter is an electrical instrument which consists of a glass electrode (a special measuring probe) attached to an electronic meter which measures the pH reading and displays it on a screen. pH is a measure of the hydrogen ion concentration (acidity or alkalinity) of a substance. Mathematically given by:

28

$$pH= -\log_{10} [H^+]\ldots\ldots\ldots\ldots\ldots\ldots (1)$$

The acidity or alkanity of a soil is measured in terms of hydrogen ion activity of the soil water system which surrounds the thin—walled glass electrode at the tip. It produces small electrical voltage of about 0.06 volt per pH unit and this enables the pH meter to measure and display the pH as unit per meter. Thus, pH of a soil is a measure of only the intensity of the activity and not the amount of the acid present. Measured on a logarithmic scale, a soil at pH 4 is 10 times more acidic than a soil at pH 5 and 100 times more acidic than a soil at pH 6. Alkalinity is usually an inherent characteristic of soils, although irrigation can increase the alkalinity of saline soils. The link between soil pH and heavy metal threshold values reflects the complex interaction between heavy metals and the various soil properties (Gawlik and Bidoglio, 2006).

3.4.2: Conductivity Measurement

This is the materials ability to conduct an electric current. In soil, it is generally used as a measure of the mineral or other ionic concentration. Electrical conductivity may also be defined as the total amount of dissolved ions in the soil and is used to estimate the amount of total dissolved salts in soil. Only ionizable materials contribute to conductivity. Its S.I unit is Siemens per meter (S/m), conductivity, σ, is given by:

$$\sigma = 1/\rho\ldots\ldots\ldots (2),$$

i.e the inverse of resistivity; which is the ratio of the electric field to the density of the current it creates, $\rho = E/J$, where, ρ =resistivity of conductor material in the Ohm meter

(Ωm), E=magnitude of the electric field (in volt per meter, Vm^{-1}), J = magnitude of the current density in Amperes per square meter (Am^{-2}).

The cause of conduction is that the metal consists of a lattice of atoms, each an outer shell of electrons which freely dissociate from their parent atoms and travel through lattice. This is known as positive ionic lattice. This "sea" of dissociable electrons allows the metal to conduct electric current. When an electric potential difference (a voltage) is applied across the metal, the resulting electric field causes electrons to move from one end of the conductors to the other. For this research, the conductivity of the soil samples was determined for each sample using a conductimeter

3.4.3: Atomic Absorption Spectrophotometry

Spectrophotometry is a quantitative measurement of transmission or reflection on properties of a material as a function of wavelength. It is more specific than the general term electromagnetic spectroscopy in that spectrophotometers deals with visible light, near-ultraviolet, and near-infrared, but does not cover time-resolved spectroscopic techniques. (Schwedt, G., 1997).

A spectrophotometer is normally used for the measurement of transmittance or reflectance of solutions, transparent or opaque solids, such as polished glass, or gases. However, they can also be designed to measure the diffusivity on any of the listed light ranges that usually cover around 200nm-2500nm using different controls and calibrations. Within these ranges of light, calibrations are needed on the machine using standards that vary in type depending on the wavelength of the photometric determination (Schwedt, G., 1997).

Of the numerous spectrophotometers Atomic Absorption Spectroscopy is considered to be the most efficient.

Atomic Absorption Spectroscopy (AAS) is used principally for the quantitative determination of metal elements in aqueous and solid samples from a wide range of fields including medicine, food and geology. The technique is an Australian invention that has its origins in the CSIRO in the 1950's and since then has become well established in laboratories around the world. **Atomic absorption spectroscopy (AAS)** is a spectroanalytical procedure for the quantitative determination of chemical elements using the absorption of optical radiation (light) by free atoms in the gaseous state.

In analytical chemistry the technique is used for determining the concentration of a particular element (the analyte) in a sample to be analyzed. AAS can be used to determine over 70 different elements in solution or directly in solid samples used in pharmacology, biophysics and toxicology research.

3.4.4: Principles of spectrophotometer

The technique makes use of absorption spectrometry to assess the concentration of an analyte in a sample. It requires standards with known analyte content to establish the relation between the measured absorbance and the analyte concentration and relies therefore on the Beer-Lambert Law; which states that there is a logarithmic dependence between the transmission (or transmissivity),T, of light through a substance and the product of the absorption coefficient of a substance, α, and the distance the light travels through the material (i.e the path length), l.

The absorption coefficient can, in turn, be written as a product of the molar concentration c of absorbing species in the material, or an absorption cross section, σ, and the (number) density N of absorbers.

For liquids, these relations are usually written as:

$$T = \frac{1}{I_0} = 10^{-\alpha l} = 10^{-\epsilon cl} \quad \text{..............(3)}$$

I_0 and I are the intensity (or power) of the incident light and the transmitted light, respectively.

Converting one to other, using $\alpha' = \alpha \ln(10) = 2.303\alpha$................... (4)

The transmission (or transmitivity) is expressed in terms of an absorbance which, for liquids is given by:

$A = -\log_{10}(I/I_0)$.............. (5) (George, R, 1976).

In short, the electrons of the atoms in the atomizer can be promoted to higher orbitals (excited state) for a short period of time (nanoseconds) by absorbing a defined quantity of energy (radiation of a given wavelength). This amount of energy, i.e., wavelength, is specific to a particular electron transition in a particular element. In general, each wavelength corresponds to only one element, and the width of an absorption line is only of the order of a few picometers (pm), which gives the technique its elemental selectivity. The radiation flux without a sample and with a sample in the atomizer is measured using a detector, and the ratio between the two values (the absorbance) is converted to analyte concentration or mass using the Beer-Lambert Law.

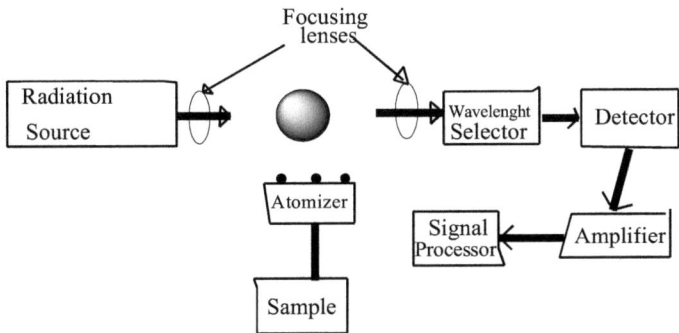

Fig.2   Atomic Absorption Spectrometer Block Diagram

In order to analyze a sample for its atomic constituents, it has to be atomized. The atomizers most commonly used nowadays are flames and electrothermal (graphite tube) atomizers. The atoms should then be irradiated by optical radiation, and the radiation source could be an element-specific line radiation source or a continuum radiation source. The radiation then passes through a monochromator in order to separate the element-specific radiation from any other radiation emitted by the radiation source, which is finally measured by a detector.

3.5: Determination of concentration of metals in soil and fruit samples

The concentration of Pb and Zn were determined in soil, cashew fruit (anacardium occidentale) and mango fruit (Mangifera indica L) samples using Flame Atomic Absorption Spectrophotometer (FAAS) Perkin Elmer Equipment model 800. The operating conditions for Pb and Zn analysis by FAAS were done with a range of standard solution with calibrated samples tested into the AAS instrument according to standard method proposed by APHP, 1998.

3.6: Method Detection Limit (MDL)

It is the minimum concentration of analyte that can be identified, measured and reported with 99% confidence that the analyte concentration is greater than zero, and is determined from analysis of a sample in a given matrix containing the analyte (USEPA, 1997).

3.7: Preparation of calibration curve using standard compounds

Standard solutions were prepared for each element depending upon the linear working range, corresponding five dilutions were made and their concentrations were measured. Standard dilutions, for each metal, was prepared from their respective stock solutions (1000 ppm) which is available readymade or prepared from their respective salts. Calibration curves were plotted using standard operating procedure.

CHAPTER FOUR

RESULTS

4.1. Physicochemical Analysis of the Soil Samples

The soil samples collected at Mokaba rehabilitated mined out site with different depths were analysed for P^H and conductivity values as shown in table 3

Table 3. pH and conductivity results.

Sampling site	pH	Conductivity (S/cm)
A	4.90	0.02
B	5.24	0.01
C	5.12	0.01
D	5.14	0.01
E	5.15	0.01
Mean value	5.11	0.01

The WHO and E.U recommended limits for pH and conductivity in soil is given in the below table

Table 4. Recommended values for pH and Conductivity by WHO and EU

Parameter	WHO	EU
pH	6-5	6-8
Electrical Conductivity (µS/cm)	250	250

4.1.2: Heavy Metal Concentration in soil and Fruits

The concentrations of the metals in the soil and fruit samples (in mg/L) from the rehabilitated mined out site of Mokaba, are given in table 5. The elements Zn and Pb have been determined by Flame Atomic Absorption Spectrophotometers (FAAS), whereas, Cr, Cu and As have been determined by Atomic Absorption Spectrophotometer.

Table 5. Concentration of dissolved metals in soil and accumulation in fruits at Mokaba rehabilitated mined site (in mg/L) dry weight.

Element	Sample		
	Soil	Mango fruit (Mangifera indica L)	Cashew fruit (Anacardium) (Occidentale)
Pb	0.053	0.355	0.055
Zn	0.222	0.396	0.685
Cr	0.027	0.019	0.025
Cu	0.817	0.847	1.140
As	ND	ND	ND

Note ND= Not Detected

Recommended limits of investigated metals (Pb, Zn, Cr, Cu and As by World Health Organization (WHO) in mg/L are as shown in table 6.

Table 6. Recommended limits of investigated metals (Pb, Zn, Cr, Cu and As by World Health Organization (WHO) in mg/L.

Parameter	WHO Recommended Permissible Limits
Lead (Pb)	0.050
Zinc (Zn)	<5.000
Chromium (Cr)	<0.050
Copper (Cu)	<1.000
Arsenic (As)	0.010

Comparison of heavy metals concentration in soil and accumulation in the fruits of Mokaba rehabilitated mined site (in mg/L) is shown in Table 7 below.

Table 7. Comparison of heavy metals concentration in soil and accumulation in the fruits of Mokaba rehabilitated mined site (in mg/L).

Elements	Samples			WHO (mg/L)
	Soil (mg/L)	Mango fruit (Mangifera indica L) (mg/L)	Cashew fruit (anacardium) (Occidentale)(mg/L)	
Pb	0.053	0.355	0.055	0.050
Zn	0.222	0.396	0.685	<5.00
Cr	0.027	0.019	0.025	<0.050
Cu	0.817	0.847	1.140	<1.00
As	ND	ND	ND	0.010

4.2: Calculations

4.2.1: Translocation Factor (TF)

The transfer capability of heavy metals from soil to the edible part of fruits was generally described using the translocation factor (I. Tsafe et al., 2012). Translocation factors (TF) of heavy metals were calculated as follows:

$$TF = \frac{\text{Metal concentration in edible part of fruit}}{\text{Metal concentration in soil from where the fruit was grown}} \quad \ldots\ldots\ldots\ldots (6)$$

Table 8: Heavy Metals (HMs) Concentration in Soil (CS), Accumulation in Mango Fruit (AMF), Accumulation in Cashew Fruit (ACF) and their Translocation Factors (TF) in mango fruit and cashew fruit respectively

HMs	CS (mg/L)	AMF (mg/L)	ACF (mg/L)	TF in Mango fruit	TF in Cashew fruit
Pb	0.053	0.355	0.055	6.698	1.038
Zn	0.222	0.396	0.685	1.784	3.086
Cr	0.027	0.019	0.025	0.704	0.926
Cu	0.817	0.847	1.140	1.037	1.395
As	ND	ND	ND	–	–

This can be diagrammatically represented as shown in figures 3 and 4

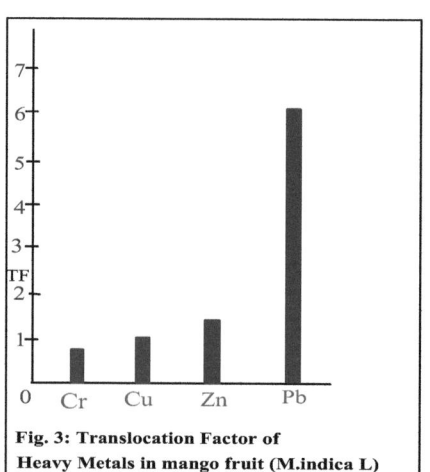

Fig. 3: Translocation Factor of
Heavy Metals in mango fruit (M.indica L)

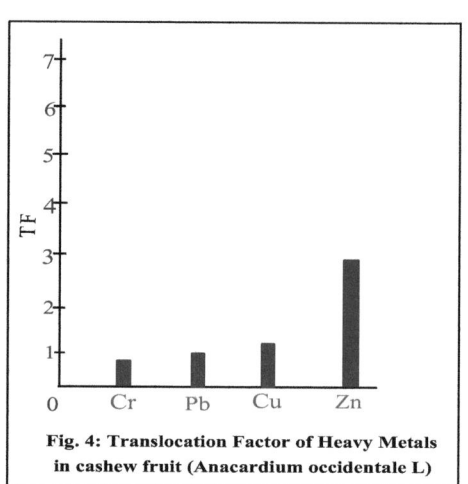

Fig. 4: Translocation Factor of Heavy Metals
in cashew fruit (Anacardium occidentale L)

CHAPTER FIVE

DISCUSSION OF RESULTS

5.1: pH and Conductivity analysis in the Soil samples

pH has a great impact on bioavailability in the soil. The lower the pH indicates an increase in metal bioavailability. From the result obtained in table 3, the Mokaba rehabilitated mined out land was found to have a pH range between 4.90—5.24 which indicates that the soil is acidic and out of range with W.H.O average values.

Electrical conductivity (EC) as a measure of the ionic concentration, or total amount of the dissolved ions in the soi.The conductivity of the Mokaba rehabilitated mined out land ranged from 0.01—0.02S/cm. l. This indicates that the land has trace metal ions or ionizable materials that will contribute to conductivity, again exceed the W.H.O average values.

5.2: Determination of Heavy Metals Concentration in the Soil and in the Fruits

In the present study, it was found that the tested dry weight peeled fruits showed higher content for some elements (especially copper and lead), while for others, the content was lower, Cu went from 0.817mg/L in soil, 0.847mg/L in mango fruit (Mangifera indica L) to above detection limit (1.140mg/L) which is as a result of bioaccumulation. Lead and copper were found to exceed the WHO recommended limit. Zn and Cu were found to be at a much higher concentration in the cashew fruit than in the mango fruit. This could be due to the age of the tree bearing the fruit. The highest Zn and Cu contents are found in the young plants. Arsenic was not detected. The results were compared with suitable safety standards recommended by W.H.O and the levels of Zn and Cr in cashew fruit (Anacardium Occidedntale L) peel dry weight and mango fruit (Mangifera indica L) were within the acceptable limits for human consumption.

The content of toxic metals (Pb and Cu) in the tested fruits was found to be significantly reasonable. As in both soil and fruits was not detected. The concentration of traces of the heavy metals Cu, Zn, Cr and Pb in the sample of soil and in mango fruit (Mangifera indica L) and cashew fruit (anacardium occidentale L) are reported in Table 8. The results showed an irregular pattern of the heavy metal availability. Cu concentration has the highest value in the soil (0.817mg/L) and also the highest in the fruits. The concentration of heavy metals in the soil follows the order: Cu > Zn > Pb> Cr whereas the concentration of metals in the Anacardiaceae and Anacardium Occidentale fruits is in the order of Cu>Zn>Pb>Cr.

Cr concentration is least among all other metal in the fruits and it is within the safety limit. These variations of heavy metal in soil and fruits could be due to differences in the sources of the metal fraction.

5.3: Translocation Factor (TF) Between Soil and Fruits

From eqn. (6):

$$TF = \frac{\text{Metal concentration in edible part of fruit}}{\text{Metal concentration in soil from where the fruit was grown}} \quad \text{............ (6)}$$

Using table 8, it is found that Pb has the highest TF value (6.698) in Mango (Mangifera indica L) and trends of heavy metals follows order Pb>Zn>Cu>Cr, and in Cashew fruit (Anacardium occidentale) the trend is as Zn>Cu>BP>Cr . It is diagramatically representated (see figs.4 and 5). This could be attributed to low retention rate of the metal in soil.

5.4: Correlation Coefficient (r^2) Between Soil and Fruits

Correlation (r) measures the relationship between two variables say x and y using mathematical formulae. For a product moment correlation coefficient, it is given by:

$$r_{xy} = \frac{N\Sigma xy - \Sigma x \Sigma y}{\sqrt{\left[N\Sigma x^2 - (\Sigma x)^2\right]\left[N\Sigma y^2 - (\Sigma y)^2\right]}} \quad \text{................. (7)}$$

N= no. of observations, r = correlation between x and y, Σ = summation.

Correlation coefficient results normally range from 0—1 negative or positive and are interpreted as:

+1 = this means there exists a perfect relationship between x and y and are in the same direction.

-1 = this means there exists a perfect relationship between x and y and in opposite direction,

0.0= this means no relationship exists between x and y,

0.8—0.9= this means there exists a very strong relationship between x and y,

0.6—7= strong relationship between x and y,

0.1—0.5= weak relationship between x and y and vice-versa

The coefficient of determination (r^2) transforms a correlation coefficient into a statistic that can be readily interpreted in terms of percentage and compared to other coefficients.

The correlation coefficients between the soil and mango fruit (**r_{sm}**) and that of soil and cashew fruit (**rsc**) were calculated using equation (7) from tables 9 and 10 below respectively.

Table 9. Estimation of correlation coefficient between soil and mango fruit.

Heavy metal	Soil (s) (mg/l)	Mango (m) (mg/l)	s^2	m^2	sm
BP	0.0530	0.3550	0.0028	0.1260	0.0188
Zn	0.2220	0.3960	0.0490	0.1570	0.0879
Cr	0.0270	0.0190	0.0007	0.0004	0.0005
Cu	0.8170	0.8470	0.6670	0.7170	0.6920
Total	$\sum s$=1.1190	$\sum m$ =1.6170	Σs^2= 0.719	Σm^2 = 1.0004	Σsm= 0.7992

From eqn. (7); letting x=s, y=m and N=4

r_{sm} = 0.925,   i.e. the correlation coefficient between the soil and the mango fruit; from which the coefficient of determination (r_{sm}^2) in terms of percentage = 85.6%.

Also, considering the relationship between heavy metal concentration in the soil and in cashew fruit (rsc) as shown in table 10

Table 10. Estimation of correlation coefficient between soil and cashew fruit.

Heavy metal	Soil (s) (mg/l)	Cashew (c) (mg/l)	s^2	c^2	sc
BP	0.0530	0.0550	0.0028	0.0030	0.0029
Zn	0.2220	0.6850	0.0490	0.4690	0.1520
Cr	0.0270	0.0250	0.0007	0.0006	0.0007
Cu	0.8170	1.1400	0.6670	1.2990	0.9310
Total	Σs=1.1190	Σc= 1.9050	Σs^2= 0.719	Σc^2= 1.7720	Σsc= 1.0860

Using eqn. (7); letting x=s, y= c and N=4.

rsc = 0.933 i.e. the correlation coefficient between the soil and cashew fruit, and the coefficient of determination (r_{sc}^2) in terms of percentage = 87%.

Hence, from the correlation analysed, there is a very strong correlation between soil and mango ($\mathbf{r}_{sm}^2$) and between soil and cashew ($\mathbf{r}_{sc}^2$) fruits. This implies that the source of the extra metal concentrated by fruits is most likely the soil.

CHAPTER SIX

CONCLUSION

Appropriate precautions should also be taken by environmental agencies at the time of mining and explorations of the soil in the Sierra Rutile environs. Cashew (Anacardium Occidentale L) and mango (Mangifera indica L) fruits showed the extremely high accumulation tendency towards the heavy metals (i.e, Cu and Pb). The species cashew fruit (Anacarium Occidentale) and mango fruit (Mangifera indica L) are found to accumulate the metals: Cu, and Pb at high levels and Zn and Cr at moderate levels, As was not detected. This study reveals that the soil in the rehabilitated Mokaba land carry levels of heavy metals. The buildup of heavy metals in soil profile may prove detrimental not only to plants and animals but also to consumers of the harvested crops from the farms.

It is therefore suggested that economic plants like rubber and anacacia are most suited in the Mokaba rehabilitated land rather than food crops.

REFRENCES

1. Anderson, M. C., J. F. Ferguson and J. Gavis., 1976. Arsenate adsorption on amorphous aluminum hydroxide. J. Colloid Interface Sci. 54:391-399.

2. Ansari, T.M., Marr, I.L., & Tariq, N., 2004. Heavy Metals in Marine Pollution Perspective. A Mini Review. J. Appl. Sci. 4(1): 1-20.

3. Aziz, H.A., Yusoff, M.S., Adlan, M.N., Adnan, N.H, Alias, S., 2004: Physico-chemical removal of iron from semi-aerobic landfill leachate by limestone filter. Waste Manag. 24: 353-358.

4. Bala M, Shehu R A,Lawal M. Determination of the Level of Some Heavy Metals in Water Collected From Two Pollution-Prone Irrigation Areas Around Kano Metropolis. Bajopas., 2008.

5. Bartlett, R. J. and J. M. Kimble., 1976. Behavior of chromium in soils: II. hexavalent forms. J. Environ. Qual. 5:383- 386.

6. Brown. G.I.,Introduction to inorganic chemistry ,1983,3rd Edition, Longman, Brown .G.I., Introduction to physical chemistry ,1983,3rd Edition,414., Longman ,Hill.G.Cand Holman,J.S.,Chemistry in context,1980-292,E.L.B.S.

7. Catlett, K.M., Heil, D.M., Lindsay, W.L., Ebinger, M.H., 2002: Effects of soil chemical properties on $Zinc^{2+}$ activity in 18 Colorado soils. Soil Sci. Soc. Am. J. 66: 1182-1189.

8. Cavallaro, N. and M. B. McBride., 1978. Copper and cadmium adsorption characteristics of selected acid and calcareous soils. Soil Sci. Soc. Am. J. 42:550-556.

9. Cooper, EM. Sims, JT. Cunningham, SD. Huang, J.W. and Berti, WR., 1999. Chelate-assisted phytoextraction of lead from contaminated soils. J Environ Quality, 28: 1709-1719.

10. Dube L, Granry JC. The Therapeutic Use of Magnesium in Anesthesiology, Intensive Care and Emergency Medicine: A Review. Can J Anesth., 2003.

11. Ekpete OA, Kpee F,Amadi Jc,Rotimi Rb. Adsorption of Chromium (Iv) and Zinc(Ii) Ions on the Skin of Orange Peels (Citrus Sinensis) J Nepal Chem Soc., 2010.

12. El-Ghawi, U.M., Bejey, M.M., Al-Fakhri, S.M., Al-Sadeq, A.A., Doubali, K.K., 2005. Analysis of Libyan Arable Soils by Means of Thermal and Epithermal NAA, Arabian J. Sci. Eng. 30; No. 1a, 147-154.

13. Elvingson, P., Agren C., 2004: Air and the environment / Per Elvingson and Christer.

14. Facchinelli, a., sacchi, e., mallen, l., 2001. Multivariate statistical and gis-based approach to identify heavy metal sources in soils. Environmental pollution 114, 313–324.

15. FAO/WHO Joint FAO/WHO food standards programme codex committee on contaminants in foods, fifth session. 2011. 64-89.

16. Faruk, O., Nazim, S., Metin Kara, S., 2006: Monitoring of Cadmium And Micronutrients in Spices Commonly Consumed in Turkey. Res. J. Agric. Biolo. Sci. 2: 223-226.

17. Finkelman, R.B., 2005. Source and Health Effects of Metals and trace Elements in Our Environment: An overview in Moore, T.A., Black, A. Centeno, J.A., Harding, J.S. & Trumm, D.A. (ed), Metal contaminants in New Zealand, Resolution press, Christchurch, New Zealand: 25-46pp.

18. Gawlik, B. M., Bidoglio, G., 2006: Results of a JRC-coordinated study on background values, EUR 22265 EN, European Commission, Ispra, Italy, ISBN 92-79-02120-6, ISSN 1018-5593.

19. George, R. "Experimental methods in modern biochemistry" W.B. Saunders Company: Philadelphia, PA., 1976. pp 46-45.

20. Gildon, P.B. Tinker, A heavy metal tolerant strain of mycor-rhizal fungus. Trans. Br. Mycol. Soc. 77 (1981) 648–649.

21. Gimmler, H., Carandang, J., Boots, A., Reisberg, E., Woitke, M., 2002: Heavy metal content and distribution within a woody plant during and after seven years continuous growth on municipal solid waste (MSW) bottom slag rich in heavy metals. Appl Bot. 76: 203-217.

22. Griffin, R. A. and N. F. Shimp., 1978. Attenuation of pollutants in municipal landfill leachate by clay minerals. EPA-600/ 2-78-157.

23. Hickey, M. G. and J. A. Kittrick., 1984. Chemical partitioning of cadmium, copper, nickel, and zinc in soils and sediments containing high levels of heavy metals. J. Environ. Qual. 13:372-376.

24. Ioannou, A., Tolner, L., Dimirkou, A., Füleky, Gy., 2003: Copper adsorption on bentonite and soil as affected by pH. Bulletin of the Szent István University, Gödöllo Hungary. pp. 74-84. Hingston, F. J., A. M. Posner, and J. P. Quick., 1971. Competitive adsorption of negatively charged ligands on oxide surfaces. Faraday Soc. 52:334-342.

25. Jarup, L... Hazards of Heavy Metals Contamination. Br. Med. Bull., 68: 167-182., 2003; Ata. S., F. Moore And S. Modabberi,. Heavy Metal Contamination and Distribution In The Shiraz Industrial Complex Zone Soil, South Shiraz, Iran. World. App. Sci. J., 6(3): 413-425., 2009.

26. Jones C., Jacobsen, J., 2003: Nutrient management module No. 7. Testing and fertilizer recommendations.

27. Lindsay, W.L., 1979: Chemical Equilibria in Soils. Wiley-Interscience. John Wiley and Sons, New York. 449 p.

28. M. Sal jasir, a. Shaker, and m.a. Khaliq, "deposition of heavy metals on green leafy vegetables sold on road sides of riyadh city,saudi arabia", bulletin of environmental contamination and toxicology, vol 75,no,5,pp,1020-1027., 2005.

29. Mahdavian SE, Somashekar RK. "Heavy Metals and Safety of Fresh Fruits in Bangalore City, India-A Case Study". J Sci Engg Tech., 2008.

30. Maleki, zarasvand., 2008. Heavy metals in selected edible vegetables and estimation of their daily intake in sanandaj, iran. Southeast asian j trop med public health 39(2):335–340

31. Mamboya, F.A., 2007: Heavy metal contamination and toxicity: studies of macroalgae from the Tanzanian coast. Ph.D Dissertation. Stockholm University. ISBN 91-7155-3746.

32. Martinez, C. E., Motto, H. L., 2000: Solubility of lead, zinc and copper added to mineral soils. Environ Pollu. 107: 153-158.

33. McLean, J.E., Bledsoe, and B.E., 1992: Behaviour of metals is soils. USEPA Ground Water Issue, EPA/540/S-92/018.

34. Miller GD, Jarvis JK, Mcbean LD. The Importance of Meeting Calcium Needs with Foods. J Am Coll Nutr., 2001.

35. Moore, P.A., Jr., Daniel, T.C., Gilmour, J.T., Shreve. B.R., Edwards, D.R., Wood, and B.H., 1998: Decreasing metal runoff from poultry litter with aluminum sulfate. J. Environ. Qual. 27: 92-99.

36. Murray, K.S., Cauvet, D., Lybeer, M., Thomas, J.C., 1999: Particle size and chemical control of heavy metals in bed sediment from the Rouge River, southeast Michigan. Environ. Sci. Technol. 33: 987-992.

37. N. Sridhara Chary, C.T. Kamala, D. Samuel Suman Raj, Assessing risk of heavy metals from consuming food grown on sewage irrigated soils and food chain transfer, Ecotoxicol. Environ.Safe. 69(2008) 513-524.

38. Nabulo, G., 2006. Assessment of heavy metal contamination of food crops and vegetables grown in and around Kampala city, Uganda. Ph.D. Dissertation, Makerere University.

39. Nabulo, G., Oryem Origa, H., Nasinyama, G.W., Cole, D. (2008): Assessment of Zn, Cu, Pb and Ni contamination in wetland soils and plants in the Lake Victoria basin. Int. J. Environ. Sci. Tech. 5 (1), 65-74.

40. Negra, C., Ross, D.S, Lanzirotti, A., 2005a: Soil manganese oxides and trace metals: competitive sorption and microfocused synchrotron X-ray fluorescence mapping, Soil Sci. Soc. Am. J. 69 (2), 353-361.

41. O'Brien, J., 2008: What are heavy metals. Plant nutrition newsletter. April Vol. 9, No. 3.

42. Oscarson, D. W., P. M. Huang, W. K. Liaw and U. T. Hammer., 1983. Kinetics of oxidation of arsenite by various manganese dioxides. Soil Sci. Soc. Am. J. 47:644-648.

43. Rai, D., B. M. Sass and D. A. Moore., 1987. Chromium (III) hydrolysis constants and solubility of chromium (III) hydroxide. Inorg. Chem. 26:345-349.

44. Roberts J.D., and CASERIO M.C. : Basic principles of organic chemistry., 1964, 365, Benjamin Linstromberg.W.W.: organic chemistry –A brief course ,1974,3rd Edition,126-127, Health.

45. Schwedt, G., 1997. The essential Guide to Analytical Chemistry. (Brooks Harderline, trans.), Chichester, NY: Wiley. (Original published 1943). pp16-17

46. Sharma, R.K., Agrawal, M., Marshall, F.M., 2004: Effects of waste water irrigation on heavy metal accumulation in soil and plants. Paper presented at a National Seminar, Bangalore University, Bangalore, Abst. no. 7, p. 8

47. Shuman, L.M., 1991: Chemical forms of micronutrients in soils. p. 113–144. In J.J. Mortvedt et al. (ed.) Micronutrients in agriculture. SSSA Book Ser. 4. SSSA, Madison, WI.

48. Slagle, A., Skousen, J., Bhumbla, D., Sencindiver, J., McDonald L., 2004: Trace Element Concentrations of Three Soils in Central Appalachia, Soil Sur. Hor... 45. No. 3.

49. Sobukola OP,Adeniran OM,Odedairo AA,Kajihausa OE. "Heavy Metal Levels of Some Fruits And Leafy Vegetables From Selected Markets In Lagos, Nigeria". Afr J Food Sci., 2010.

50. Sresty TVS, Rao KVM., 1999. Ultra structural alternation in response to zinc and nickel stress in the root cells of pea. J env exp bot 41:3–13.

51. Tahir, N.M., Chee, P.S., Maisarah, J., 2007: Determination of heavy metals content in soils and indoor dusts from nurseries in dungun, terengganu. The Malaysian J. Analy. Sci. Vol 11, No 1. 280-286.

52. The Royal Society of Chemistry, Aspects of lead pollution,7, June ,1982.

53. U.S. EPA, 2003: Supplemental guidance for developing soil screening levels for superfund sites. U.S. Environmental protection agency, office of solid waste and emergency response, Washington, D.C.

54. USEPA\., 1997. Guidelines establishing test procedures for analysis of pollutants (App. B, Part 136, Definition and procedures for the determination of the method detection limit): U.S. Code of Federal Regulations, Title 40, revised July 1. 265–267pp.

55. WHO., 1992. Cadmium. Environmental health criteria, geneva. Vol., 134.

56. Wierzbicka, M., 1999. The effect of lead on the cell cycle in the root meristem of Allium cepa L. Protoplasma. 207:186-194.

57. Woolhouse, H. W., 1993: Toxicity and tolerance in the response of plants to metals, pp.

58. Woolson, E. A., 1977a. Fate of arsenicals in different environmental substrate. Environ. Health Perpect. 19:73- 81.

59. Zarchara, J. M., C. E. Cowan, R. L. Schmidt and C. C. Ainsworth., 1988. Chromate adsorption on kaolinite. Clays Clay Miner. 36:317-326.

60. Zhang, M.K., He, Z.L., Stoffella, P.J., Calvert, D.V., Yang, X.E., Xia, Y.P., Wilson, S.B., 2004: Solubility of phosphorus and heavy metals in potting media amended with yard waste-biosolids compost. J. Environ. Qual. 33: 373-379.

LIST OF TABLES

Table 1. Fruit samples analyzed.. 3

Table 2. Typical total concentration of Fe, Cu, Zn and Mn in soil.....................23

Table 3. pH and conductivity results.. 35

Table 4. Recommended values for pH and Conductivity by WHO and EU.......... 36

Table 5 Concentration of dissolved metals in soil and accumulation in fruits at Mokaba rehabilitated mined site (in mg/L) dry weight.............................…........….37

Table 6. Recommended limits of investigated metals (Pb, Zn, Cr, Cu and As by World Health Organization (WHO) in mg/L...38

Table 7. Comparison of heavy metals concentration in soil and accumulation in the fruits of Mokaba rehabilitated mined site (in mg/L)....................................39

Table 8: Heavy Metals (HMs) Concentration in Soil (CS), Accumulation in Mango Fruit (AMF), Accumulation in Cashew Fruit (ACF) and their Translocation Factors (TF) in mango fruit and cashew fruit respectively.....................................…......40

Table 9. Estimation of correlation coefficient between soil and mango fruit.........44

Table 10. Estimation of correlation coefficient between soil and cashew fruit.......45

LIST OF FIGURES

Figure 1. A Sketched Map of the Sample Collection Site............................. 26

Figure 2. Atomic Absorption Spectrometer Block Diagram.......................... 33

Figure 3. Translocation Factor of Heavy Metals from Soil to Mango Fruit..........40

Figure 4. Translocation Factor of Heavy Metals from Soil to the Cashew Fruit..... 40

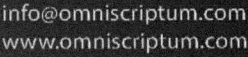

Printed by Books on Demand GmbH, Norderstedt / Germany